Animal

by Carol Pugliano-Martin

I like acorns.
I get acorns from trees.

What am I?

I am a squirrel.
I can climb trees.

I like seeds.
I get seeds
from sunflowers.

What am I?

I am a bird.
I can fly
to the sunflowers.

I like leaves.
I get leaves from trees.

What am I?

I am a deer.
I can eat leaves
from the trees.

I like flies.
I get flies in my web.

What am I?

I am a spider.